PICTURE LIBRARY

OIL RIGS

OIL RIGS

R. J. Stephen

Franklin Watts

London   New York   Sydney   Toronto

© 1986 Franklin Watts Ltd

First published in Great Britain
    1986 by
Franklin Watts Ltd
12a Golden Square
London W1R 4BA

First published in the USA by
Franklin Watts Inc
387 Park Avenue South
New York
N.Y. 10016

First published in Australia by
Franklin Watts
14 Mars Road
Lane Cove
2066 NSW

UK ISBN: 0 86313 417 3
US ISBN: 0-531-10185-1
Library of Congress Catalog Card
Number 85-52091

Printed in Italy
by Tipolitografia G. Canale & C. S.p.A. - Turin

Designed by
Barrett & Willard

Photographs by
Amoco
BP
Conoco
Chris Fox
Shell
Statoil

Illustration by
Rhoda and Robert Burns

Technical Consultant
Chris Fox

Series Editor
N. S. Barrett

Contents

Introduction

Oil rigs are used to drill for oil. Oil, or petroleum, provides nearly half the energy used in the world. Fuels made from oil power aircraft, ships, cars, trucks, trains, cranes, earthmovers and farm equipment. They are also used to generate heat and electricity.

Most oil is found deep underground. Oil rigs drill for oil on land and at sea.

△ An oil rig and platform. The tall derrick handles sections of drill pipe. This type of rig is towed by tugs to drilling sites. The platform, the yellow structure under the derrick, remains and continues to produce oil.

The search for oil is carried out by oil rigs. Once an offshore oilfield has been proved by exploration drilling, fixed platforms are generally used to produce the oil.

Depending on the size or type of field, offshore production platforms either use wells drilled by mobile drilling rigs or drill their own. Often, a combination of both methods is used.

△ Workmen connecting sections of drill pipe. Drilling is a tough job, especially in extreme climates. These vary from the icy land of Alaska and the cold, stormy waters of the North Sea to the burning deserts of Saudi Arabia.

The exploration rig

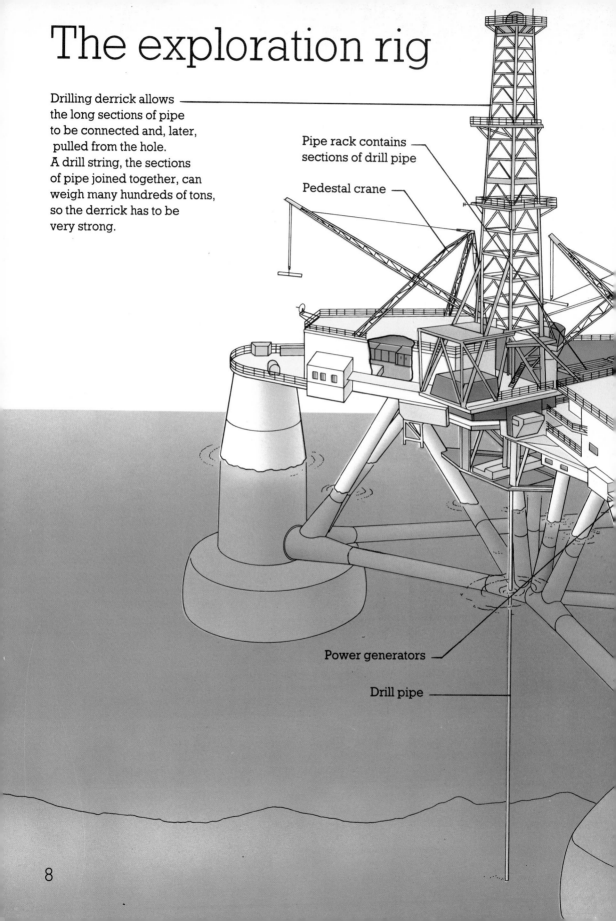

Drilling derrick allows the long sections of pipe to be connected and, later, pulled from the hole. A drill string, the sections of pipe joined together, can weigh many hundreds of tons, so the derrick has to be very strong.

Pipe rack contains sections of drill pipe

Pedestal crane

Power generators

Drill pipe

Helicopter for landing
supplies and workers

Helicopter landing deck

Living quarters

Containers

Casing for drilled well

Anchor winder

Water pumped in and out
to lower and raise rig

Pontoon

Ballast tanks contain water

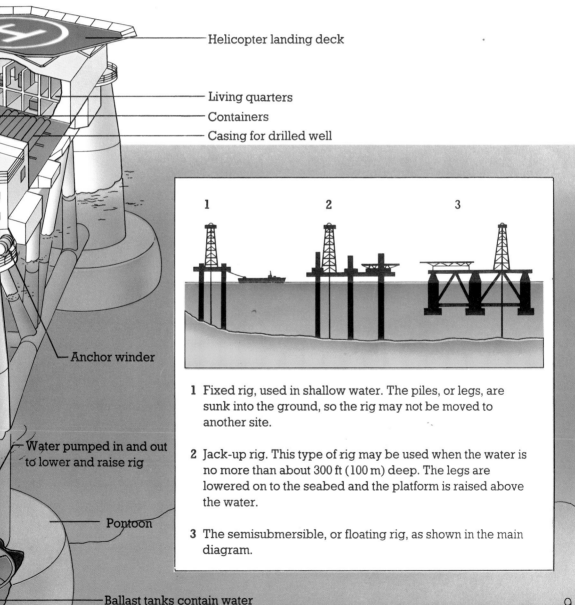

1 Fixed rig, used in shallow water. The piles, or legs, are sunk into the ground, so the rig may not be moved to another site.

2 Jack-up rig. This type of rig may be used when the water is no more than about 300 ft (100 m) deep. The legs are lowered on to the seabed and the platform is raised above the water.

3 The semisubmersible, or floating rig, as shown in the main diagram.

Kinds of oil rigs

Oil rigs are moved around, even on land. After drilling a well, a rig is dismantled and taken to another location for drilling.

The type of rig used for offshore drilling depends on the depth of the water and the conditions.

▷ A simple land rig and, on the right, two offshore jack-up platforms.

▽ A drill ship is kept in position by computers and anchors. The drill pipe goes through a hole in the middle of the ship.

Drilling for oil

The large bits used for drilling for oil work a little like a carpenter's drill. They bore out the earth as they turn around.

The drill bit is attached to the end of a long pipe. The pipe is rotated by a turntable on the floor of the derrick.

▽ The drill crew at work on the floor of an oil rig. These workmen are called driller's helpers. The one on the left is operating the tongs, which is like a big wrench. It is used to tighten or loosen the joints connecting sections of drill pipe.

The drill pipe is made up of a series of pipes. The first section of pipe is connected to the drill bit. When the top of this section gets down to the drill floor, the crew attaches another section, or joint, about 30 ft (10 m) in length.

Some offshore wells are drilled down for 10,000 ft (3,000 m) or more, and need hundreds of joints.

△ A workman (left) takes a sample of mud. This is a thick liquid that is pumped down the drill pipe to cool the bit and support the hole. It returns to the surface between the pipe and the sides of the hole.

A workman (above) handles a section of casing from the derrick. He is tethered to a strut by a safety chain.

Land rigs

Oil has been found under all the continents of the world. It lies beneath deserts and farmland, in mountainous areas and under the frozen lands of North America.

The drilling rig is transported to the site, usually by truck. The crew sets up the derrick and assembles the drilling equipment, pumps and tanks.

Oil is found under all kinds of terrain. Many tests may be made before starting a well.

△ An oil rig on the hillside farmland of Turkey.

▷ An oil rig among the sand dunes of Dubai, in the Middle East. Transport can be difficult over the soft sand of desert lands.

Some drilling sites need special preparation. The land may have to be bulldozed flat before the rig can be set up. In parts of Alaska, gravel, wood and sometimes even ice are used to strengthen the site. Otherwise, the heat of the drilling might melt the frozen earth and the well would collapse.

△ Winter drilling in the north of Canada. The icy regions of North America and the Soviet Union provide some of the harshest conditions for onshore drilling.

▷ The system of valves that controls the flow of oil at the well head is called a "Christmas tree."

▽ The pump used to help oil to the surface where it does not come up naturally is called a grasshopper or horse head.

Offshore drilling

Drilling for oil usually costs much more at sea than on land, chiefly because of the huge rigs needed. Offshore drilling is also more dangerous. In some waters, storms and icebergs are extra hazards.

Offshore rigs require a base to drill from. In shallow waters, a jack-up rig is used, with its legs resting on the seabed. Semisubmersible rigs or drill ships, which float, are used in deeper water.

△ A platform in the North Sea. Some offshore platforms are constructed in separate sections for reasons of safety or convenience. The helicopter landing-deck is on the left, above the living quarters. The large structure is the drilling platform and the other sections handle supplies.

▷ Most offshore
drilling platforms are
like this one, with
everything built as part
of one structure.

◁ Some offshore
drilling takes place on
land. This man-made
island is built up with
gravel over very
shallow water.

▷ Helicopters provide
a "bus" service for an
offshore oilfield. They
carry passengers and
supplies between the oil
platforms and shore.

The flames and
smoke that fill the air
come from the flare
stacks. Most oil wells
produce gases along
with the oil and the flare
stacks are for safely
burning off the excess
gas. However, in many
cases the gas, too, is
collected and
transported ashore for
sale.

Offshore rigs are built on land and towed out to sea. The parts of platforms are usually floated out and assembled on site by cranes.

To get the best from an oilfield, several wells are drilled from the same platform. Instead of being drilled straight down, the wells spread out over a wide area. More than 40 wells may be drilled from a single platform.

△ A semisubmersible drilling rig in choppy waters.

▷ The major parts of oil platforms are built onshore and then floated out to sea. The left-hand column of pictures shows a concrete platform being built and then towed out to the oilfield by tugs. The right-hand column shows a steel-jacket platform being built and then floated out.

△ A helicopter brings a relief crew to work on a drilling platform.

◁ A member of the drilling crew prepares a new drill bit. A used bit is on the right. The metal bits have wheels with sharp teeth which bore through the ground. The teeth are often diamond-coated. But they soon get blunt when drilling through hard rock.

Working on an offshore oil rig is hard, especially in places where it is cold and stormy, such as the Arctic Ocean and the North Sea. The drilling crew operates under very difficult conditions. The crew has to contend with the wind and rain and the continual noise of the drilling. But it is usually a well-paid job.

Most companies have two crews working 12-hour shifts, with another two crews relieving them after 7 or 14 days.

△ Life on oil platforms is tough, and the drilling crew and other workers might spend several days or even weeks on these "islands" in the sea. Their leisure hours are made as comfortable as possible. The meals are usually of a high standard and there are game rooms and other recreation areas.

Supply and maintenance

Offshore rigs and platforms need regular servicing. Supplies and changes of crew and other workers have to be brought in by supply ships and helicopters, often in difficult conditions.

Divers do all kinds of inspection and repair work below water. They make shallow dives in a special suit. For deeper work, a team of divers operates from a diving bell.

▽ A supply ship bringing in chemicals to a production platform. Supply vessels are tough, sturdy boats, able to withstand gale force winds and rough seas. Unloading a supply ship or landing a helicopter in bad weather can be a very tricky operation.

△ *Stadive*, on the left, is a semisubmersible service vessel designed for both emergency and routine operations. It is used for fire-fighting and can support as many as 24 divers.

▷ A diver uses an electronic instrument to check the underwater structure of an oil platform. Divers must be skilled workers, able to use many kinds of tools and inspection devices.

The story of oil rigs

△ Signal Hill, California, in the 1920s, showing the bustle of early oilfields.

△ Drilling in the Middle East in the 1920s, in what is now Iran.

The first oil rig

In ancient times, people used oil to make a kind of tar, or bitumen. They collected the oil that seeped to the surface from underground springs. In the mid-1700s, oil appeared from wells drilled for salt. It was at first regarded as a nuisance.

In the 1850s, scientists discovered methods of producing kerosene from the crude oil. Kerosene was used for lighting lamps. The oil was brought up in buckets from wells dug by workmen. In 1859, Edwin Drake built the first oil rig, near Titusville, Pennsylvania. Using a wooden rig and a steam-operated drill, he struck oil at a depth of 69 ft (21 m).

The oil boom

Soon, thousands of oil wells were drilled in the Pennsylvania

hills. Towns sprang up all over the region. The oil was transported to factories on the coast to be refined, first by wagon and river barge, then by rail and pipeline. The first pipeline was built in 1865.

Other American states also began to produce oil and Texas started large-scale production in 1901. By that time, the total American production had increased from 2,000 barrels a year in 1859 to 64 million barrels.

The gasoline engine

The gasoline engine was invented in the late 1800s. Gasoline had been produced from crude oil along with kerosene, but it exploded if burned in a lamp. Refineries dumped the unwanted product. This changed dramatically with the development of gasoline-

28

driven cars. Mass production of cars in the early 1900s led to an increased demand for crude oil, and gasoline soon became the major fuel produced from it.

Now, crude oil also provides kerosene for aircraft and diesel fuel for ships, trucks, trains and other vehicles and machines. Other important oil products include paraffin for heating, bitumen for road-building and chemicals for plastics and paints.

The Middle East

The countries of the Middle East produce more than a third of the world's oil. The first oil from the region was found by a British company in Iran in the early 1900s. Since then, American, British and other world powers have helped to develop oilfields throughout the Middle East,

which is largely desert. Special vehicles were designed for transporting oil rigs across the sands.

Offshore oil

The first offshore oil wells were sunk just off the coast in oilfields that had already been developed onshore. Offshore drilling began as early as 1891 off the coast of California. The wells were drilled from structures linked to the land, like seaside piers. Offshore fields were later developed in Venezuela and the Soviet Union.

The first mobile offshore rig was built in 1949, for use in shallow water. A few years later, floating rigs were developed and a jack-up rig was used in the Gulf of Mexico. Semisubmersible rigs were built in the 1960s to explore for oil under the North Sea.

△ Drilling in the Middle East in the 1980s, in Dubai.

△ A modern jack-up oil rig, for use in shallow water.

Facts and records

Giant

The BP magnus platform, in the North Sea, is the world's tallest, at a height of 1024 ft (312 m). This is a few feet taller than the Eiffel Tower in Paris. It delivers 120,000 barrels of oil a day.

△ BP Magnus is the world's tallest oil production platform.

On tow

In 1981, eight tugs towed the Statfjord B platform to its site in the North Sea from Stavanger, Norway. At 900,000 tons (816,000 mt), it was the heaviest object ever moved, weighing the equivalent of more than eight of the world's largest aircraft carriers.

△ The Statfjord B platform was the heaviest object ever moved.

Desert transport

Transporting exploration rigs across the soft desert sands in many countries of the Middle East presents its own problems. Air transport is expensive, so special trucks are used. They have huge tires that will not sink into the sand.

△ A special truck for transporting exploration rigs across soft sand.

Glossary

Barrel
An old unit of measurement, but one that is still used for oil. One barrel is equivalent to 42 gallons (160 l).

Crude oil
Oil as it comes from the ground, before being refined into special fuels and chemicals.

Derrick
The framework over the borehole, used for lowering or hoisting up sections of drill pipe.

Drill bit
The tool that cuts through the ground to make the hole.

Drill pipe
The hollow tube, made up in sections, which is turned by the rotary table on the rig floor and itself turns the drill bit.

Flare stack
A long arm at the end of which gas is burned off.

Jack-up rig
A drilling rig used when the water is not deeper than about 300 ft (100 m). The legs are lowered to the bottom when drilling.

Kerosene
One of the products of crude oil. It is used mainly as aircraft fuel.

Mud
A special mixture of chemicals that is pumped through the hollow drill pipe. It cools the bit, controls the pressure and supports the hole. It returns up the outside of the pipe with bits of rock that can be tested for oil.

Oil rig
The structure and machinery used to drill for oil. It can be moved from site to site.

Petroleum
Another term for oil.

Production platform
A structure used for producing oil from offshore wells. Large production platforms, made of steel or concrete, can also drill new wells and maintain existing ones.

Roughneck
A driller's helper.

Semi submersible rig
A rig used for deep-sea drilling. It is supported by the buoyancy of its submerged pontoons.

Index